FLUORESCENT PAPER

PROPERTIES

Materials	Printed paper
Features	Highly visible
Dimensions	Various
Supplier	Local stationery stores
Function	Village notice boards, W.I. posters and greasy spoon café signage systems
Rethink	Book jacket

RETH TOM DIXON NK

SPECIAL PHOTOGRAPHY BY ASHLEY CAMERON

conran
OCTOPUS

First published in 2000 by
Conran Octopus Limited
a part of Octopus Publishing Group
2–4 Heron Quays
London E14 4JP

www.conran-octopus.co.uk
ISBN:1 84091 256 1
This paperback edition published in 2002

British Library Cataloguing-in-Publication Data
A catalogue record for this book is available from
the British Library.

Commissioning Editor: Denny Hemming
Project Editor: Kate Bell, Hilary Burden
Editorial Assistant: Alexandra Kent

Creative Director: Leslie Harrington
Design: Johnson Banks
Stylist: Sarah Hollywood

Picture Research: Clare Limpus, Liz Boyd
Production Manager: Zöe Fawcett

Rethink is about looking at the world of made objects in a different way. Trying to find the hidden beauty in the mundane. Spotting fitness for purpose where it exists, even when not intended. Sometimes taming the industrial artefact for domestic use. But always keeping your eyes wide open.

START

START Witness the radical change of function and alignment for these standard 12.2m- (40ft-) long shipping containers, diverted to a new use (right) by Dutch architect Henk Tilder, as radical low-cost housing in Almere, north west of Amsterdam.

START The art of casual intervention turns polystyrene packing cases into a fishing boat in Azerbaijan (above), and puts unseaworthy boats back to work as poetic beach huts on Lindisfarne, Holy Island, Northumberland (top).

How confusing are the decoration, lighting and furniture trade fairs with their vast selection of overstyled products? Skim through design magazines and you find more of the same — the latest reworking of 70-year-old modernism. A fresh generation of design graduates resorts to ever-increasing contortions to try and come up with a new twist on interior decoration. But how many more shapes do we need a chair to be?

We are surrounded by extraordinary examples of natural and anonymous designs; those with fitness for purpose, radical styling, and clever innovation. But how many of us really stop to notice? Anonymous worlds of unrecognized and unnoticed designs are waiting to be discovered and acknowledged by an industry which has failed to embrace fully its ancestry. Do we really have to wait for the next World War to unclutter our heads and blast a torrent of fresh air through this stale discipline? I think the answer lies around us.

As a business, design is a relatively recent occupation. The career now described as "designer" would have traditionally been carried out by talented amateurs, engineers, industrialists and inventors as a natural by-product of their core interest: more normally to develop an object's FUNCTION rather than its IMAGE.

Let's finally pay some overdue respect to the unselfconscious, talented people who evolved the breeze block, to the Chinese seamstress who prototyped the first cuddly toy, and to the engineer who developed the ubiquitous brass stopcock that hides away in every utility room.

START Worn tyres remain the most versatile and available of rethought materials. Commonly seen on motor racetracks as safety barriers, they are also universally used in playgrounds (the example, right, is in Tokyo). In both cases, improvisation creates common design practice.

For some, the ugliest form of rubbish, for others a raw material just waiting to fulfil its potential: (above) reused tyres form a wall, or, (below) provide valuable playthings for youngsters.

We ought regularly to seek answers to obvious questions: who put the dip into a standard house brick, and who devised the size and spacing of the pattern on a piece of aluminium tread-plate? We will never know and the anonymity could be part of the attraction. Yet these people are no less worthy of our attention because their names have gone unrecorded.

Flick through the pages of any issue of Industrial Exchange & Mart, visit a military museum or walk into a plumber's shop: strange but true shapes will appeal and speak in ways that the latest fabricated trends from TV decoration programmes never do.

Such unfamiliar objects, furnishings and components are almost invisible. They just exist and remain largely unnoticed, often due to their omnipresence. The radical uses of new materials and the exploitation of new functions and typologies are happening all around us, but we just don't recognize them. All that is needed is that we start to look afresh.

What to do with these products? Use them everywhere, that's what. Use them for their intended purpose or not.

The truth is that there is nothing remotely new about this idea. Almost anybody might use an oil drum as a barbecue, and all over the world plastic carrier bags become rubbish bins. It is universally accepted in junior circles that a car inner tube makes the perfect toy boat. Meanwhile, the outer part of the same tyre is used for anything from flip flops to water containers or buffers on seagoing trawlers. The hub of a wheel is variously used as an anvil or a table base according to need. And it was not surprising to find that in the orbiting space station MIR, cosmonauts used wire coathangers as radio aerials.

If we take this logic a stage further we find that in clothing, for instance, army surplus has for years appeared as youth fashion and has recently hit the mainstream as the obvious solution for high function, low-cost, hard-wearing clothing. In industry it is routine to diagonally slice a plastic milk container and use the base for standard nut and bolt storage.

In any impoverished environment people will naturally take whatever comes to hand and adapt its use to a task. Spotting new uses for existing artefacts seems an almost natural human activity. Thus we can see the clever appropriation of the glass bottle as building brick. A smart trick indeed, as it is cheaper and allows in considerably more light than the common or garden clay house brick. The ubiquitous discarded Coke can is similarly used, but this time its structural integrity and phenomenal economy of material is used to maximum effect as a lightweight building module.

We see Indian families colonize stacks of huge irrigation ducts in an ethnic version of the Japanese capsule hotels — this improvization is obviously a product of necessity but is nonetheless symbolic of man's infinite creativity.

This primitive fishing float in Piraeus, Greece, is constructed from discarded 5 litre oil bottles, and is strangely beautiful in its simplicity. It is also curiously well resolved decoratively with its colourful repeated branding. Design or accident?

In Tripoli, Lebanon (left), recycled wine bottles are seen in a new guise as a free building material for roof domes, showing man's urge to make do and improvize. In the Hamman al Jadid (above), light is let in, while the interior is kept cool: a perfect example of beauty meeting purpose.

Bricks are easy to consider as a banal building material, yet the Minimalist Carl André was almost universally condemned for claiming them as fine art. Yet "Equivalent VIII", 1966, is a striking piece. Damien Hirst also uses everyday objects, such as the medical cabinet (below), "Whisper", 1997, to create art installations.

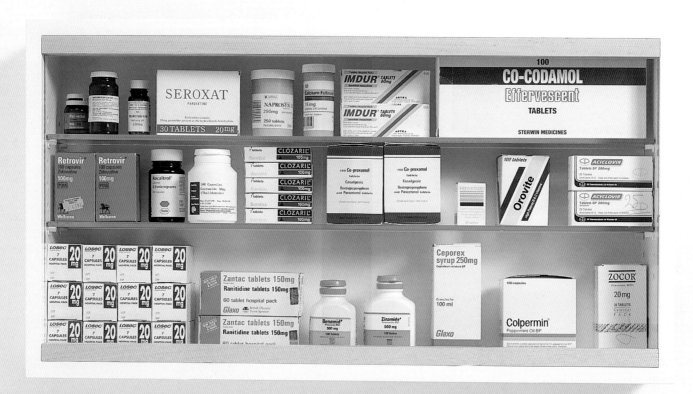

But this book is not simply about customization, misappropriation or man's ingenious ability to recycle materials into surprising new functions. It is more about encouraging oneself to think about surrounding artefacts in a new way. In one sense it is about celebrating what we have but have so far failed to notice. In another, it is about noticing beauty and functionality in mundane materials or the everyday industrial artefact and then transporting them to the domestic environment. Not because it is trendy or wacky, but because it is economic and effective.

Twentieth-century artists have often touched on and understood the mysterious power of the industrial object: Marcel Duchamp, the French painter and sculptor, whose seminal artwork made from a public pissoir reminds us just how many lives an object can have; Picasso, who could always see other forms arising from the most banal objects (a bicycle saddle as a bull's head being the obvious example); Joseph Beuys, whose love-hate relationship with industrial felt made for the most disturbing and monumental artworks; and the British artist, Damien Hirst, whose installations made from surgical instruments and medical cabinets have a chilling precision.

Designer Rody Graumans, for Droog Design, works with, and finds, the beauty in mundane industrial and domestic items. This light bulb chandelier, "85 Lamps", 1993, is remarkable, made solely by the multiplication of wire bulbs and electrical connector blocks.

The disposable coathanger from the drycleaners is much too good to throw away. Even birds think so. Reappropriation is not always solely a human activity — as witnessed in the extraordinary image of a nesting crow. What is it that possessed this bird to filch metal coathangers to build a colour-coded nest? Could it be the same biological principle that makes children need toys, no matter how impoverished?

Industrial, agricultural and nautical catalogues display intriguing and practical objects, but what precisely is so appealing about the awkward but unselfconscious forms that appear on their pages? The naive but workmanlike way that they are presented is usually compelling, photographed as crude cut-outs or against a plain white backdrop without props or shadows. A self-conscious office junior is often photographed with them to demonstrate its function or scale. The manufacturers are seeking to maximize the information that can be squeezed onto a page.

The repetitive nature of the cataloguing certainly lends a rhythm to these publications, as does the multiple options and configurations offered. The bright colours, often chosen for safety or coding reasons, are also very striking. Can we, with a fresh set of values and our minds focused on a new aesthetic, look at these objects in a different light? Certainly the whole office sector is riddled with ugly furniture often produced for no reason other than its lowest possible production cost.

Most can be discarded for their complete and utter ugliness. However, through a careful and rigorous selection, can we use them to our decorative advantage? Is it possible, if we all believe it hard enough, that woodgrain teak effect mahogany will come back into fashion?

The most engaging thing about all of these artefacts is their pure functionality, their fitness for purpose. They seem to have evolved rather than to have been designed. This is the true form follows function that the modernists strove for, and it is available to all who care to visit trade counters, or send off for industry-only catalogues.

The ubiquitous road sign formed into sculpture on a roundabout. At night, the signs reflect vehicle headlights. This "Landmark" by Pierre Vivant, 1992, was commissioned by the Cardiff Bay Arts Trust.

START Great pleasure is to be found by those who adapt existing technologies or objects to new functions. Why make yet another style and shape of chair when when things for other uses can do the job just as well?

POLYSTYRENE CUP

PROPERTIES

Materials	Insulated, expanded polystyrene
Features	Heat proof
Capacity	207 or 295ml
Supplier	ABE Catering Equipment
Function	Disposable cup
Rethink	Chandelier

LIVE

LIVE   Rethink? It's like this. Take a new look at your living space. Abandon any cravings for fashion. No need to spend until you've looked, and looked again at what's there already. And 'there' can be anywhere: your kitchen, the local café or council workyard, even an office supplier.

CARPET TILES

PROPERTIES

Materials	Burmatex Velour Plus — tough polypropylene and nylon
Features	Anti-static, resistant to stains; shampoo to clean
Dimensions	500 x 500mm
Supplier	Key Industrial & Commercial Equipment
Function	Ideal for offices, contract applications, exhibitions, hospitals, libraries, etc.
Rethink	Domestic high-traffic areas

There are certainly advantages to using a contract office equipment catalogue as a starting point for a domestic sitting room. The fierce competition in this market sector guarantees the lowest possible price and home delivery is the norm. Hard-wearing finishes are standard, and decorative trimmings are eschewed, giving this range of products a very basic, low-grade minimalism which has an appeal all of its own.

In our living room we find the ubiquitous and much maligned carpet tile, reviled for its synthetic nature and coarse pile. Here, however, it is cleverly introduced to the domestic environment in a daring selection of colours, reminiscent of a designer rug. Superiority as a flooring material soon becomes apparent, with its ease of transport, simplicity of installation and hard-wearing nature. It also has the added blessing of being extremely economic.

The polystyrene coffee cups used to construct the large central light fitting were simply bonded together with a glue gun. No other component is required to make this spectacular chandelier. It is both lightweight and affordable — the only cost is the time.

CASTOR

PROPERTIES

Materials	Medium/light-duty, pressed steel swivel; polypropylene and nylon wheel
Features	Braked castor, using foot-operated pedal
Capacity	Loading capacity 125kg
Supplier	Key Industrial & Commercial Equipment
Function	Suitable for most applications
Rethink	Coffee table leg

An industrial pallet loader with massive castor wheels and a textured rubber top is used as a coffee table. The flower arrangement is simply a wholesale tray packed with bulk hyacinth bulbs. Why replant? Display them in their original box instead.

A second-hand depot supplied the modular reception area sofa which is easily reconfigured according to specific entertaining needs. Originally intended for office reception areas, it provides a flexible solution to the complicated nature of modern home entertaining. Perfect for TV dinners.

A modern twist on wallpaper: advertising posters, destined for an exterior hoarding, can been used inside as low-cost pop art.
As they tend to be printed in over-ambitious quantities, try obtaining one free of charge through the relevant hoarding site company.

LIVE Students are often adept at rethinking their environment because meagre budgets force them to make do. Using bricks as furniture is not new, then, but making a virtue of improvization is. Here, an extravagant screen is created using the basic breeze block.

The amazing variety of brick and building block types worldwide remains undocumented. In Mali, for example, bricks are fabricated from cow dung, clay and straw which are compressed into moulds and left to dry in the sun — rather than baked — giving a pleasing, irregular appearance.

Meanwhile, the French prefer a lightweight, extruded, hollow brick, which seems unavailable in the UK. Instead, the type favoured here has changed from being the rectangular, earthenware shape, to a more solid grey breeze block — made for more rapid and cheap construction.

Although seemingly less attractive than the standard house brick, they make ideal Lego-style bricks for temporary interior structures.

BREEZE BLOCK

PROPERTIES

Materials	Cement
Features	Light and inexpensive
Dimensions	Various
Supplier	Any builders' merchants
Function	Low-cost walling material
Rethink	Adult construction kit

Stack breeze blocks (cheap, lightweight, and usable without mortar) into simple monolithic forms to create Flintstone-like coffee tables (above) or even sofas (top). The dust problem is easily overcome by painting with a matt varnish.

LIVE The clever slatted construction of these security doors (above) makes them flexible but almost indestructible — a perfect solution for temporary walls or screens (right). Available in many versions including galvanised steel and extruded aluminium.

Other misuses? Cut to table height, they make a handy base for a wooden- or glass-topped table. In the garden, cut to 30cm (1ft) in length and sunk partly into the earth, they can be used as a minimal, curvaceous restraining wall to allow a drop in levels.

For their own purposes, theatrical suppliers or packaging specialists process large rolls of thin, flexible plastic film into 2 x1 metre semi-rigid panels. These can be generated with any surface pattern, ranging from brickwork effect to tudor half timbering. With a little persuasion, the manufacturer can often be persuaded to produce the plastic in transparent forms, or, as in the example shown (left), metallic gold. This has been vac-formed into an alpine rock face for a kitsch Liberace-in-Las Vegas feel. Panels have the great advantage of being extremely cheap, lightweight, and easily fixed into position with a staple gun to cover crumbling walls or ceilings, or to make temporary walls.

Car tyres are one of the most offensive forms of dumped rubbish known to man. All over the world, huge landfill sites are dedicated to their disposal. Their very indestructability, however, gives them limitless uses beyond their original function: as crash barriers on motor racetracks, or as swings in adventure playgrounds, sustaining battering by generations of kids.

They also contain within them an even more exciting and adaptable resource — the latex inner tube. Cut into strips, it makes excellent seating, webbing, or bungees for the car roof rack, and in its original state of inflation, provides unconventional seating at no cost.

Second-hand inner tubes are generally given away for free by garages.

INNER TUBE

PROPERTIES

Materials	Latex
Features	Inexpensive, re-use
Dimensions	Various
Supplier	Tyre supplies
Function	Tyre inner tube
Rethink	Seating

Kids do it. Bring the playground inside by introducing a sense of fun into your adult living space. Take comfort in a truck tyre inner tube.

HAZARD LAMP

PROPERTIES

Materials	Manmade
Features	Two directional with flashing amber light; supplied in packs of 5
Dimensions	330 x 80mm
Supplier	Signs & Labels
Function	Road danger lamp
Rethink	Decorative lamp

CARTONS

PROPERTIES

Materials	Recyclable cardboard
Features	Flat-packed for easy assembly
Dimensions	Choice of 6 medium and 3 heavy-duty sizes
Supplier	Key Industrial & Commercial Equipment
Function	Heavy-duty cartons for international freight
Rethink	Partition blocks

WARNING MAT

PROPERTIES

Material	3mm rubber with grooved surfaces for slip resistance
Features	Textured backing; available in a variety of bright colours
Dimensions	610 x 864mm
Supplier	Seton Identification & Safety Buyer's Guide
Function	Safety signmat
Rethink	Doormat

SHELVING

PROPERTIES

Materials	Powder-coated finished steel; chipboard shelving
Features	No nuts, bolts or back bracing required; frames and beams clip together
Dimensions	Versatile choice of shelving dimensions
Supplier	Key Industrial & Commercial Equipment
Function	Riv-Rak warehouse shelving system built to accommodate most industrial, commercial and retail storage requirements
Rethink	Bookshelf

TELEPHONE BOOTH

PROPERTIES

Materials	Glass fibre reinforced plastic lined with 25mm thick acoustic foam
Features	Sound-proofed, wall-mounted and available in four colours
Dimensions	H790 x W740 x D580mm
Supplier	Slingsby Commercial & Industrial Equipment
Function	Privacy for wall-mounted payphone
Rethink	Lounge phone booth

TRAY TROLLEY

PROPERTIES

Materials	High-strength extruded aluminium with a natural finish
Features	Supplied in knock-down form and quick and easy to assemble using 12 bolts; choice of slide spacing available
Dimensions	H1630 x W520 x D710mm
Supplier	Slingsby Commercial & Industrial Equipment, and Mailbox
Function	Ideal for catering applications, front or back of house
Rethink	TV and video unit/CD store

STACKABLE CHAIRS

PROPERTIES

Materials	Textured polypropylene with chromed steel frame
Features	Low-noise feet; easily washable, suitable for indoor and outdoor use; built-in link to connect chairs together in rows
Dimensions	Seat 430 x 410mm; Back 400mm
Supplier	Bernard
Function	Office or school chairs
Rethink	Dining chairs

STORAGE UNIT

PROPERTIES

Materials	3mm mild steel, corrugated for extra strength
Features	Sloping roofs, galvanized grid mesh removable floors, double steel sliding doors
Dimensions	H2800 x W5600 x DH1500mm
Supplier	Slingsby Commercial & Industrial Equipment
Function	Drum storage to help companies comply with environmental pollution regulations
Rethink	Walk-in wardrobe

WHEELY BIN

PROPERTIES

Materials	Aluminium
Features	Cut out side for ease of loading; push handles on both ends
Capacity	945 litres
Supplier	Slingsby Commercial & Industrial Equipment
Function	Giant transport truck
Rethink	Sofa base

TRAFFIC LAMP

PROPERTIES

Materials	Heavy gauge steel with stove epoxy finish for all-weather operation
Features	Flashing rate: 120/150 per minute; approx battery life 1200–1500 hours
Dimensions	H250 x W170 x D76mm
Supplier	Key Safety & Hygiene Equipment
Function	Road traffic safety and security
Rethink	Party light

PARK WASTE BIN

PROPERTIES

Materials	22 gauge galvanized sheet steel
Features	Epoxy coated; easy to empty by swivelling; can be fixed to a wall or post; rectangular or cylindrical models
Capacity	27 litres
Supplier	Bernard
Function	Public waste bin
Rethink	Domestic waste bin

VACUUM CLEANER

PROPERTIES

Features	Lightweight, 1000W motor
Capacity	9 litre dry capacity
Supplier	Bernard
Function	Retail/office cleaner
Rethink	Domestic vacuum

CHAIR

PROPERTIES

Materials	Vinyl upholstered seat; deep radius polypropylene back
Features	Black with chrome finish; foldable; handy moulded hand hold in chair back for ease of handling;
Dimensions	H889 x W457 x D446mm
Supplier	Key Industrial & Commercial Equipment
Function	Chair for offices, colleges, schools, etc
Rethink	Domestic use

TOOL TROLLEY

PROPERTIES

Materials	All steel construction
Features	Extremely sturdy, easy to manoeuvre; supplied flat-packed with simple assembly instructions
Dimensions	W1357(open) 873(closed) x D535 x H792mm
Supplier	Slingsby Commercial & Industrial Equipment
Function	Mobile tool trolley
Rethink	TV or Hi-fi console

STEEL SHELVING

PROPERTIES

Materials	High quality steel, bright zinc plated and laquered
Features	Airdeck design to allow flow of air for fresh and clean storage; adjustable shelf levels
Dimensions	1700mm high bays; comes in four widths and three depths
Supplier	Key Industrial & Commercial Equipment
Function	Used in medical environments and the electronics industry
Rethink	Living-room shelving

NEON SIGN

PROPERTIES

Materials	Glass signage encased in clear acrylic sheeting
Features	For interior use only; DIY installation; free-standing or hanging
Dimensions	Various
Supplier	Signs & Labels
Function	For attracting passers-by and creating a welcome environment
Rethink	Artpiece

KIK-STEP STOOL

PROPERTIES

Materials	Tough steel contruction
Features	Non-slip rubber pads and retracting castors
Dimensions	H355mm
Supplier	Signs & Labels
Function	For offices, libraries and schools
Rethink	Home stool or foot rest

HALOGEN LIGHT

PROPERTIES

Materials	Rubber-encased steel and brass with plastic lamp head
Features	50W halogen worklights provide bright and focused light; adjustable through 360° in both vertical and horizontal planes
Dimensions	500 or 700mm long lamp arm; lamp head 100mm long x 52mm diameter
Supplier	Key Industrial & Commercial Equipment
Function	Work lights
Rethink	Bedside or desk lamp

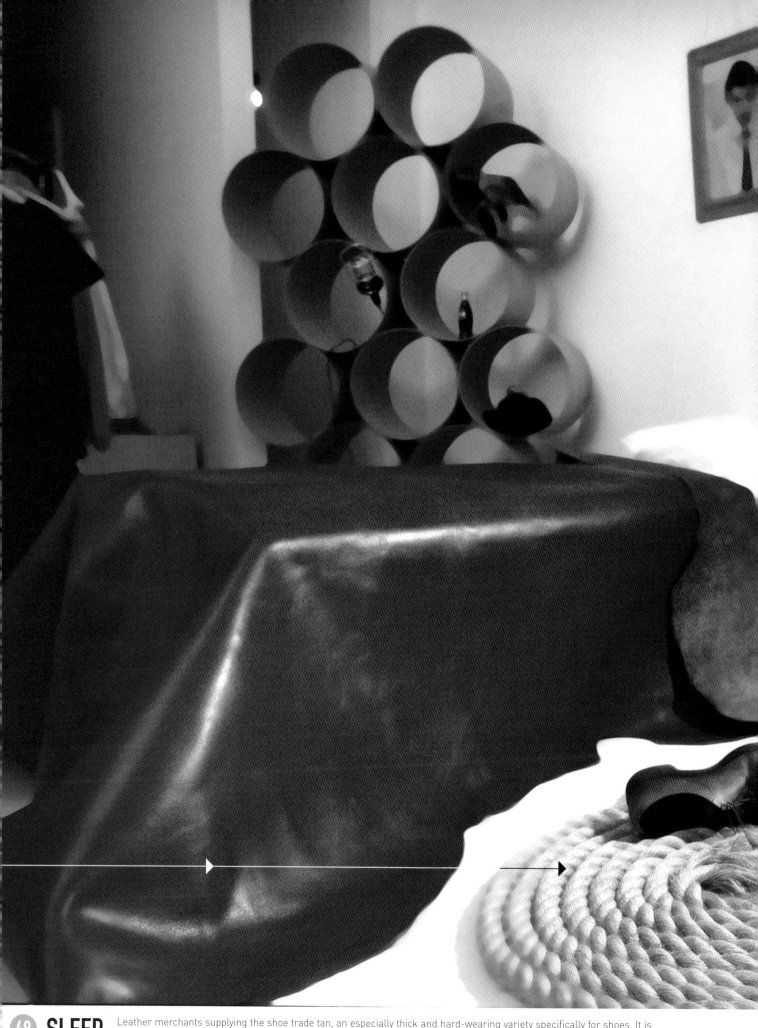

SLEEP Leather merchants supplying the shoe trade tan, an especially thick and hard-wearing variety specifically for shoes. It is tough enough to use as flooring or work surfaces, but in this case has been used rather flamboyantly as a bedspread.

Large section hemp rope can still be found in thick coils at yacht chandlers where it is still favoured over the nylon variety. It can be used for a variety of non-traditional functions. Shown here, it is worked into an incredibly hard-wearing circular rug.

A wide selection of utilitarian lamps is used in industry and tends to be more versatile and more durable than anything available in decoration shops.

This halogen site lamp is a good example. It comes with two heads which can pivot in any direction, and is fully adjustable in height. It is usable inside or out and is easily transportable as the tripod is retractable. Toughened glass lenses which protect the tungsten halogen strips are themselves protected by wire grilles. The ensemble is powder coated in a tough epoxy finish in safety yellow.

BUNSEN BURNER

PROPERTIES

Materials	Metal
Features	Self-contained and fully portable with gas cartridge – convenient where no gas supply
Capacity	220g butane/propane cartridge allows minimum burning time (4hrs approx).
Supplier	Philip Harris Scientific
Function	Laboratory bunsen burner
Rethink	Table light

SLEEP Large cardboard tubes, used in the transportation of all kinds of goods, are usually discarded. Remarkably tough and resistant to compressive forces, they can be used to make anything from table legs to small buildings.

Large diameter tubes, sliced and bolted together to form an
attractive room divider, also serve as shelving and storage space.

Pallets are often abandoned when cargo reaches its destination. But consider using them as decking, or a bed base. They are available in wood and plastic, and in a wide variety of sizes.

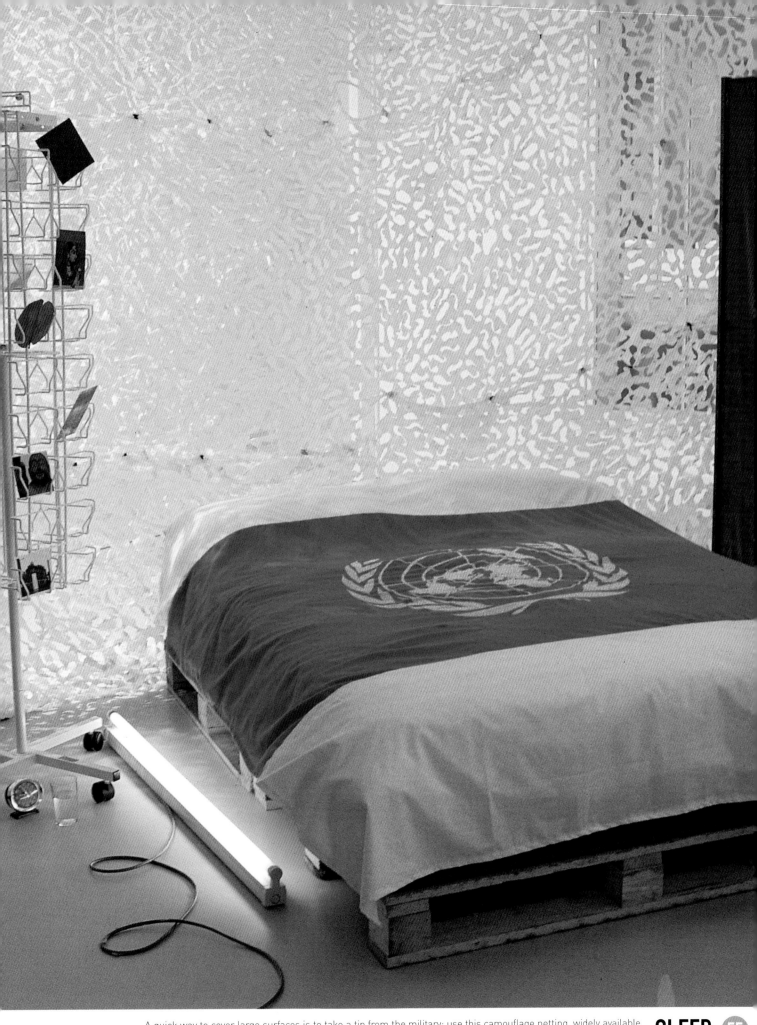

A quick way to cover large surfaces is to take a tip from the military: use this camouflage netting, widely available from army surplus stores in several colours, to suit various combat situations — desert, snow or forest.

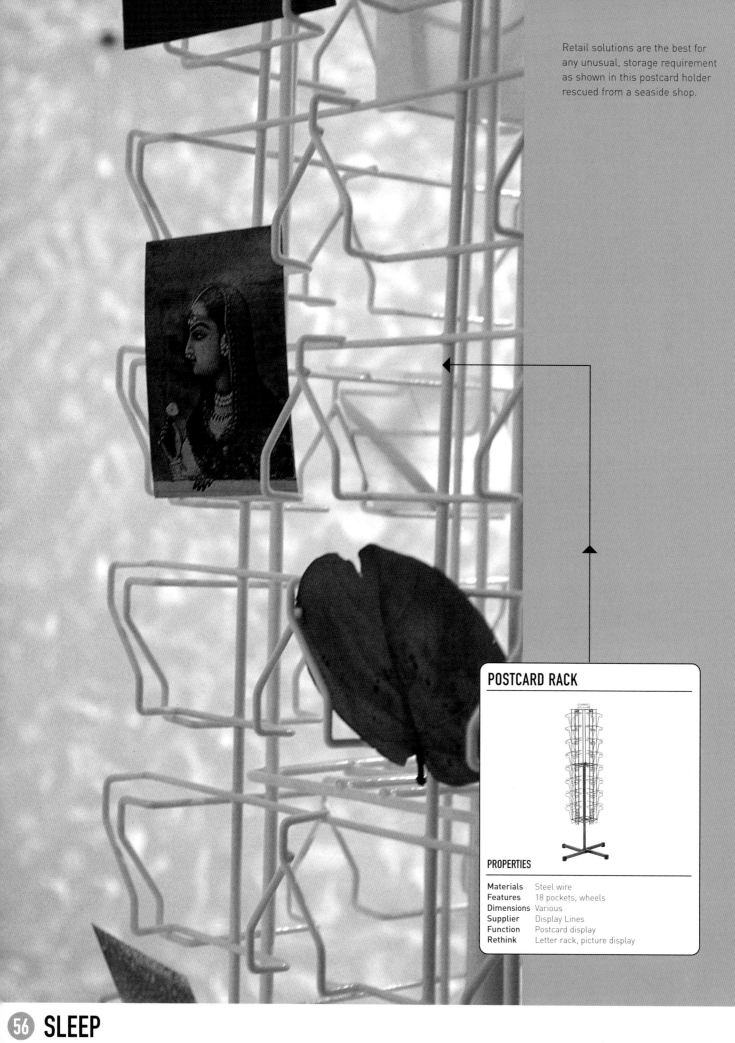

Retail solutions are the best for any unusual, storage requirement as shown in this postcard holder rescued from a seaside shop.

POSTCARD RACK

PROPERTIES

Materials	Steel wire
Features	18 pockets, wheels
Dimensions	Various
Supplier	Display Lines
Function	Postcard display
Rethink	Letter rack, picture display

The commercial electronic light display, often seen in banks and airports, is used here as a bedroom clock. It could also work as a message board in a home office. Instant art and the time on your wall.

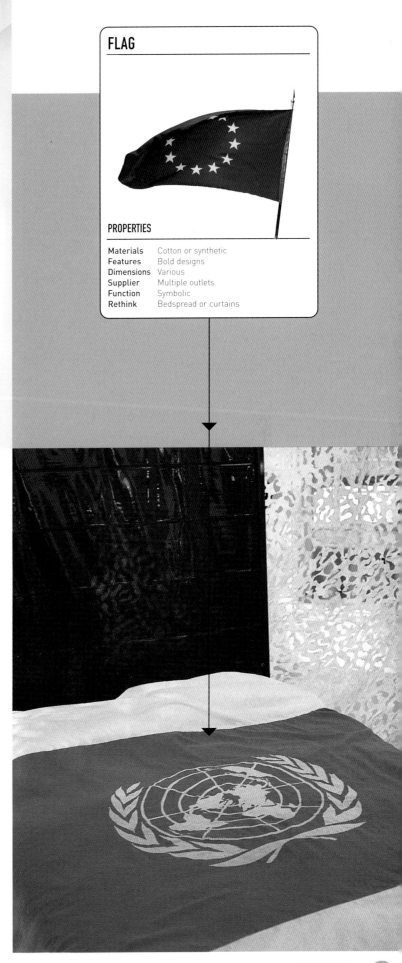

FLAG

PROPERTIES

Materials	Cotton or synthetic
Features	Bold designs
Dimensions	Various
Supplier	Multiple outlets
Function	Symbolic
Rethink	Bedspread or curtains

Flags are available in a wide variety of sizes and make a perfect decorative bedspread or curtain for people who resist floral prints but still want a decorative look. The bedhead is a new use for a plastic welding shield.

MOVING DISPLAY SIGN

PROPERTIES

Features	Single and double line displays, available in a variety of colours; 8000 character memory; comes complete with programming key pad
Dimensions	H100 x W720 x D40 or H200 x W1520 x D90mm
Supplier	Signs & Labels
Function	Illuminated display to convey company message
Rethink	Artwork, domestic messages

LOCKER

PROPERTIES

Materials	Sheet steel
Features	Single, two, three or four doors; choice of four colours
Dimensions	300 or 460mm depth
Supplier	Rapid Racking
Function	School locker
Rethink	Wardrobe

BOX TRUCKS

PROPERTIES

Materials	Fibreboard
Features	Lightweight, double reinforced seams
Capacity	75, 100 and 125kg
Supplier	Key Industrial & Commercial Equipment
Function	Storage
Rethink	Bedlinen containers

GARMENT RAIL

PROPERTIES

Materials	Tubular steel
Features	Lightweight, mobile, available in three styles
Dimensions	H1160 x L1050
Supplier	Slingsby Commercial & Industrial Equipment
Function	Wholesale fashion garment rail
Rethink	Domestic clothes rail

SCREEN

PROPERTIES

Materials	PVC flame-retardant curtains
Features	Four folding panels mounted on free-running mini castors
Dimensions	H1745mm
Supplier	Slingsby Commercial & Industrial Equipment
Function	Hospital ward screen
Rethink	Bedroom screen

PREPARATION BENCH

PROPERTIES

Materials	White laminate on MDF core
Features	Preparation table with part boxes and lower shelf
Dimensions	H1390 x W2400 x D755mm
Supplier	Bristol Maid Hospital Equipment
Function	Sterile medical preparation table
Rethink	Bedside table

LOCKER

PROPERTIES

Materials	All steel
Features	6 compartments, supplied with camlock and two keys
Dimensions	H1829mm
Supplier	Slingsby Commercial & Industrial Equipment
Function	Widely used in factories, schools, hospitals
Rethink	Underwear or trinket storage

MIRROR

PROPERTIES

Materials	Shatterproof acrylic
Features	Available in both convex and hemispherical styles; will not fade or discolour
Dimensions	Available in six styles and sizes
Supplier	Seton
Function	Safety and security
Rethink	Bedroom mirror

TELEPHONE CABINET

PROPERTIES

Materials	Medium density polyethylene
Features	Available in four colours with logo, can be post or wall mounted
Dimensions	H420 x W330 x D256mm
Supplier	Slingsby Commercial & Industrial Equipment
Function	Office phone cabinet
Rethink	Home phone

CABINET

PROPERTIES

Materials	Galvanized sheet steel
Features	Removable metal shelves; door fitted with rubber sealing strip
Dimensions	H1219 x W915 x D635mm
Supplier	Pottery Crafts Ltd
Function	Damp storage cabinet for pottery
Rethink	Cupboard

LAMP

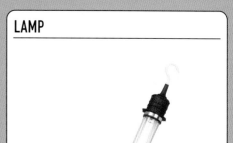

PROPERTIES

Materials	Fluorescent, insulated lamp with 6m rubber cable
Features	Water tight; oil, petrol and dust resistant
Dimensions	L520 x W55mm
Supplier	Key Industrial & Commercial Equipment
Function	Lightweight, mobile lamp
Rethink	Bedroom lighting

UTILITY TROLLEY

PROPERTIES

Materials	Mild steel-welded construction, epoxy powder coated
Features	Mounted on rubber tyred swivel castors
Dimensions	L550 x W430 x H910mm
Supplier	Slingsby Commercial & Industrial Equipment
Function	Workshop tool trolley
Rethink	Bedside table

GRIP LAMP

PROPERTIES

Materials	Nylon lamp holder; coated steel guard
Features	Adjustable reflector; 5m plastic covered cable; switchable
Capacity	100W max; bayonet fitting only
Supplier	Key Industrial & Commercial Equipment
Function	Gripper inspection lamp
Rethink	Bedside table lamp

BIN

PROPERTIES

Materials	Resin fibreboard
Features	Tapered for nesting
Dimensions	Available in three sizes
Supplier	Key Industrial & Commercial Equipment
Function	Storage and distribution container
Rethink	Linen store

PLAN FILING CABINET

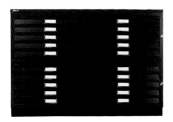

PROPERTIES

Materials	Steel with quality powder-coated finish
Features	Anti-tile features and central locking
Dimensions	Available with ten 60cm high drawers or five double height drawers
Supplier	Key Industrial & Commercial Equipment
Function	Plan file for the neat and efficient storage of a large number of drawings
Rethink	Bed base

WORKBENCH

PROPERTIES

Materials	High quality mild steel
Features	Worktop available in six options; supplied with lockable drawer, cupboard and drawer plate
Dimensions	H840 x W1200 x D900mm
Supplier	Slingsby Commercial & Industrial Equipment
Function	Heavy-duty workbench
Rethink	Dressing table

CUPBOARD

PROPERTIES

Materials	Mild steel, white epoxy powder coated
Features	Adjustable shelving, lockable doors
Dimensions	H1930 x W915 x D310mm
Supplier	Bristol Maid Hospital Equipment
Function	Pharmacy cupboard
Rethink	Bedroom cupboard

TIDY TROLLEY

PROPERTIES

Materials	Polypropylene containers
Features	Mounted on portable trolley
Dimensions	Available in eight sizes and five colours
Supplier	Slingsby Commercial & Industrial Equipment
Function	Portable trolley for easy part identification
Rethink	Underwear or trinket trolley

BARREL

PROPERTIES

Materials	Oak
Features	Natural materials, rugged construction
Dimensions	Various
Supplier	Coopers' suppliers
Function	Bulk liquid transport
Rethink	Bath

WASH

WASH The cut down brewing barrel is an ideal container for bathwater, providing a natural alternative to the standard tub shape, and is a much more convenient shape for those needing to share a bath.

The plumber's shop is a wonderful source of valves, stopcocks, piping and junctions — all fascinating components for a grown-up's playing kit. Bath taps are easily assembled and provide an economical alternative to chromed Victoriana or overstyled 'designer' hardware currently available. Towel rails are often unimaginative. Here (above) a functional ladder is draped with towels.

CORK

PROPERTIES

Materials	Cork
Features	Eco material, light, reuse
Dimensions	35 x 15mm
Supplier	Brewing supplies, wine bottles
Function	Sealing bottles
Rethink	Bath mat

WASH A practical alternative to bathroom floor tiles, paving stones are so ubiquitous that they pass unnoticed. They are, however, an invaluable resource for low-cost, hard-wearing waterproof flooring material.

WASH Toilet rolls are cumbersome to handle and bulky to store, and invariably run out when you have guests to stay. Turn toilet tissue into an attraction by buying in many colours in bulk.

Packaging suppliers for industry sell rolls of corrugated cardboard that can be used in a multitude of ways. In this instance, the material is used as a wall covering with the corrugations working externally. It is a pleasing alternative to the usual wood chip used for covering poor quality or uneven walls, and provides a padded hard-wearing surface for high traffic areas. Painting roughly with a roller creates an interesting two-tone effect as the paint will sit only on the raised edges of the corrugations.

TURF EFFECT CARPET

PROPERTIES

Materials	100% polypropylene
Features	Latex backing with drainage slits, good for damp areas
Dimensions	W2m
Supplier	Bernard
Function	Swimming pool surrounds
Rethink	Hygienic carpet

Stainless steel is a luxury option in sanitary ware, and is traditionally used in areas that are likely to suffer extreme abuse such as prisons or public conveniences. However, if you can afford them, they make the most handsome of toilets. Also available are sinks, shower units and urinals — all offering the same workmanlike simplicity.

PAPER ROLL HOLDER

PROPERTIES

Materials	Epoxy coated steel
Features	Wall-mounted holder
Dimensions	Suitable for rolls up to 400mm wide
Supplier	Bernard
Function	Wall-mounted dispenser
Rethink	Domestic toilet roll holder

FIRST AID KIT

PROPERTIES

Materials	Vinyl water-resistant zip pouch
Features	Supplied complete with contents
Dimensions	H250 x W200 x D25mm
Supplier	Key Safety & Hygiene
Function	HSE approved workplace first aid kit
Rethink	Domestic first aid kit

STEP STOOL

PROPERTIES

Materials	Injection moulded plastic
Features	Spring-loaded castors, with anti-slip studs to hold it in place
Supplier	Signs & Labels
Function	Cost-efficient mobile office step stool
Rethink	Bathroom seat

INDUSTRIAL LAUNDRY TRUCK

PROPERTIES

Materials	MPDE
Features	Available in five colours
Capacity	450 litres
Supplier	Mailbox Healthcare Products
Function	Tags and seals for medical security
Rethink	Domestic laundry basket

PLASTIC FLOOR MATTING

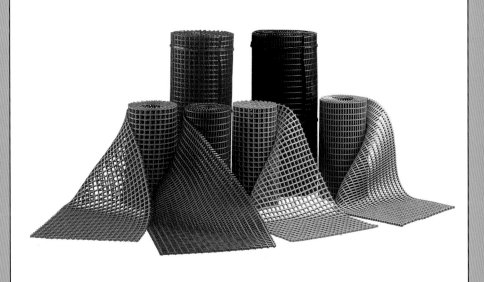

PROPERTIES

Materials	PVC
Features	Chemical resistant, flame retardant, 2 sizes of mesh
Dimensions	L5 or 10m x W600/900mm
Supplier	Welconstruct
Function	Designed to ease fatigue and provide an anti-slip working environment
Rethink	Bathroom floor covering

MEASURING JUG

PROPERTIES

Materials	High-density polyethylene
Features	Available in yellow and white; will not chip, crack, dent, discolour or corrode
Capacity	½–5 litre
Supplier	Key Industrial & Commercial Equipment
Function	Measuring liquid
Rethink	Bath jug

JANI-JACK TROLLEY

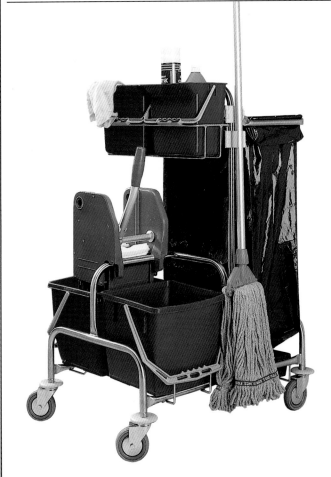

PROPERTIES

Materials	Rust-free non-magnetic stainless steel
Features	Adjustable height, mop wringer, two broom holder clips
Dimensions	H1100 x W830 x D540mm
Supplier	Bernard
Function	Used in food preparation and healthcare environments
Rethink	Bathroom mop and cleaning storage

SHOWER

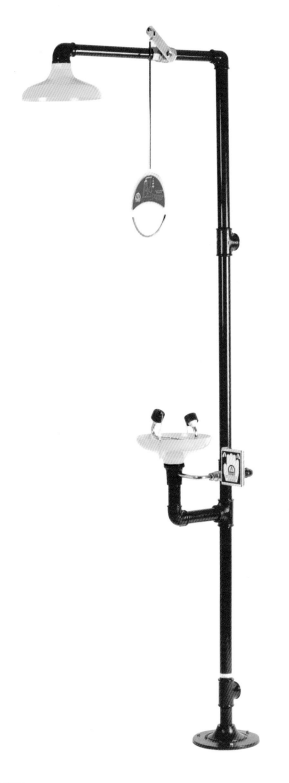

PROPERTIES

Materials	Plastic
Features	Shower head operated by large pull cord; wall or ceiling mounted
Supplier	Slingsby Commercial & Industrial Equipment
Function	Emergency first aid equipment
Rethink	Domestic shower

HAND DRYER

PROPERTIES

Materials	Impact-resistant polycarbonate cover with white polyester gloss finish
Features	Fully automatic, energy saving touch-on touch-off
Dimensions	H261 x W272 x D140mm
Supplier	Key Safety & Hygiene
Function	Public toilet hand dryer
Rethink	Wall-mounted hair dryer

SPOUT JUG

PROPERTIES

Materials	High density polyethylene
Features	Complete with screw-on, leak-proof, flexible spout
Capacity	1-5 litres
Supplier	Key Industrial & Commercial Equipment
Function	Allows liquid to be poured into apertures situated at awkward angles
Rethink	Hair rinsing jug

INDUSTRIAL SOAP DISPENSER

PROPERTIES

Materials	Wall mounted; durable plastic case with keylock
Features	Complete with wall fixing and screws, easy to replace 500ml soap cartridge (sold separately)
Dimensions	H310 x W100 x D125mm
Supplier	Bernard
Function	Washroom soap dispenser
Rethink	Bathroom soap dispenser

WASTE BINS

PROPERTIES

Material	Polypropylene
Features	Colour code guide recommended by the British Institute of Cleaning Sciences: blue for general; red = toilets and washrooms; yellow = wash-basins and showers; green = food preparation and bar use
Capacity	54 litres
Supplier	Bernard
Function	Industrial waste bin
Rethink	Laundry basket

UTILITY TROLLEY

PROPERTIES

Materials	One piece moulded structural foam, polymer
Features	Impact-resistant surface will not chip, peel or dent; rounded edges with no crevices make cleaning easy; grey or blue finish
Capacity	70kg per shelf
Supplier	Slingsby Commercial & Industrial
Function	Versatile industrial utility trolley
Rethink	Trolley for bathroom towels

BOWL STAND

PROPERTIES

Materials	Stainless steel
Features	Stands for different bowl sizes; anti-static castor wheels
Dimensions	Bowl ring diameter 350mm
Supplier	Bristol Maid
Function	Hospital equipment
Rethink	Washstand

PORTABLE TOILET

PROPERTIES

Materials	Cassette C-200
Features	Control panel with touchbutton electric flush, and an automatic signal when the holding tank is full
Supplier	Thetford
Function	Roadgoing caravan/motorhome portable loo
Rethink	Domestic lavatory

CABINET

PROPERTIES

Material	Zintec-coated mild steel, finished in white polyester
Features	Wall mounted or free standing; fitted with strong T handle
Dimensions	H610 x W458 x D152mm
Supplier	Signs & Labels
Function	Workplace safety cabinet
Rethink	Medicine cabinet

SECURITY SEALS

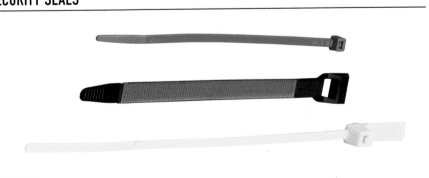

PROPERTIES

Materials	Plastic
Features	Robust security
Dimensions	Full range in varying widths and sizes
Supplier	Mailbox Healthcare Products
Function	Tags and seals for medical security
Rethink	Hanging bath toys and bathroom sundries

BENCH

PROPERTIES

Materials	Powder-coated finished steel; chipboard
Features	Easy to assemble, no nuts, bolts or bracing
Dimensions	H1574 x W1524 x D620mm
Supplier	Key Industrial & Commercial Equipment
Function	T-bar bench with drawer, louvred panel, top shelf for multi-purpose workstation
Rethink	Kitchen bench

COOK

COOK This garage workbench provides the perfect base for a flexible micro kitchen. The rear louvred panel can be accessorised with bins or hooks, for foodstuffs and implements. The work height is adjustable and the full width tool drawer gives maximum storage.

Garages, workshops and factories all need lightweight, hard-wearing, affordable, adaptable and easily transportable work surfaces. These requirements could also be specifications for an economy kitchen, which is why, as illustrated, the transition is easily made.

In some situations, the functionality of the workbench is far superior, and certainly more economical than its household equivalent. The height of the work surface is also adjustable and new components are easily bolted on. The metal finish is a hard-wearing epoxy powder coat which comes in a variety of bold colours, providing a radical alternative to the anaemic shades normally available. The bench itself is mounted on castors, making the whole unit reminiscent of a restaurant kitchen, which is obliged to be mobile for hygiene reasons, allowing for rigorous cleaning behind stoves and work surfaces where germs or pests might lurk.

INDUSTRIAL WARNING MAT

PROPERTIES

Material	3mm rubber with grooved surfaces for slip resistance
Features	Textured backing; available in a variety of bright colours
Dimensions	610 x 864mm
Supplier	Seton Identification & Safety Buyer's Guide
Function	Safety signmat
Rethink	Kitchen mat

LOUVRE PANEL BINS

PROPERTIES

Materials	Polyethylene
Features	Easy to clean, also avail. in red, blue, green
Dimensions	Various sizes
Supplier	Slingsby
Function	Small parts storage
Rethink	Container for food and cleaning stuffs

Although designed solely with function in mind, these taps have a gawky appeal all of their own. They manage to avoid the self-conscious pretence of the many taps available in hardware shops. A useful nozzle shape allows a rubber hose to be slipped over the outlet, and they are easily wiped clean.

The back panel of the unit is covered in a louvred panel — an industry standard in assembly plants or repair workshops. They can be accessorized with a wide variety of containers, which are easily removed for replenishing, or for handy positioning during the cooking process. Available in food-safe polyethylene, they are equally appropriate for consumables or hardware, cutlery or gadgets. If you prefer a more natural and less aggressively industrial feel, woodworking tables and benches provide a solid beech alternative.

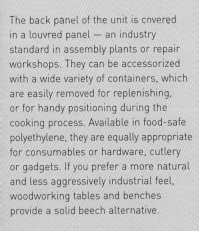

Kitchen sinks and their various accessories always seem inadequate compared with their industrial counterparts. There is never enough space or versatility in the dishrack, overstyled taps, puny little scourers and shallow basins with no room to swing a cat. Luckily, in the professional sector we find much more satisfactory products. A fully adjustable wall-mounted rack sourced from a chemical laboratory is shown opposite.

Conventional kitchens show a miserable lack of imagination in the sink and hardware departments. The choice can easily be broadened by looking at other receptacles, such as the red nursery play tray shown on the right.

LABORATORY RACK

PROPERTIES

Materials	High impact polystyrene
Features	Injection moulded with 72 holes & relevant pegs, expansion hanging hooks and draining tube
Dimensions	630 x 450mm
Supplier	Philip Harris Scientific
Function	Laboratory draining rack
Rethink	Kitchen dishrack/drainer

Supermarket shelves are filled with endless variants of cleaning products. Subtle distinctions in shape and a wide range of colours make them attractive vessels indeed, especially when stripped of their branding label. Their functionality is beyond questioning.

They can be refilled, as bulk packs are now widely available although be careful not to mix up chemicals. Or, use them as temporary vases or unusual collectibles. There is a lot to be said for giving them pride of place on the window sill while using them for their original funtion. **COOK** 85

COOK The modern domestic kitchen is not noted for its character with its easy-wipe, anti-bacterial surfaces. Above, we see a serious alternative, perhaps harder to handle than the usual compact laminate, but a constantly evolving one that will last for life and consistently improve.

A large proportion of any piece of timber is discarded to meet the universal demand for straight, flat planks of wood.
However, the rejected portion often holds the true beauty of the material with knots, bark and random dimensions.

Stripped of their negative connotations, prison plates are surprisingly functional for those who like to exercise portion control at home, and are perfect for children who like to pretend they are on an aeroplane or train. They have the added advantage of being indestructible, and provide a useful function to telly addicts determined to enjoy uninterrupted viewing — no need to stop for dessert because it can be included in a compartment at the time of serving.

Only in Western cultures is it deemed necessary for kitchens to have an oven and four gas or electric rings. Add to this the toaster, pressure cooker, microwave, food processor and ice cream machine, and kitchen areas become highly complex.

Followers of the modernists' maxim of 'form follows function' would do well to visit an army barracks to see the authentic incarnation of this theory. No fancy chrome finishes or luxury leathers are on display. There are no overworked details or expensive fittings. Everything is constructed for durability and efficiency. Fixings are simple and rugged, materials are honest and low key, and portability and compactness are essential features. The army beaker, below, is a classic example of this type of honesty in design.

Some of the most delicious cuisines in the world, however, are managed with only one compact heat source. Vietnamese cooks are expert at making a meal of dumplings steamed above a pan of soup. Moroccans manage to make an oven-like device called a Tagine over a small charcoal burner. Australians pride themselves on their prowess with the barbecue.

Why, then, does it seem a foregone conclusion that every apartment needs a full-blown fitted kitchen with an expensive cooker? The gas ring illustrated is from an Indian restaurant, and is designed to heat the round bottomed, Balti-style cooking pot popular all over Asia and available in multiple dimensions.

MOP BUCKET

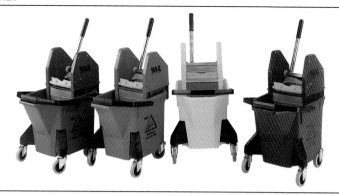

PROPERTIES

Materials	Polypropylene with non-rusting castors
Features	Strong durable wringer and pouring spout
Capacity	20 litres
Supplier	Key Industrial & Commercial Equipment
Function	Industrial floor cleaning
Rethink	Kitchen mop

INDUSTRIAL FLOORING

PROPERTIES

Materials	PVC matting
Features	High-quality, laboratory tested, hygienic, easy to maintain by brushing, easily rolled up for cleaning
Dimensions	3 widths: 600, 800 or 1000mm
Supplier	Bernard
Function	For barefoot walking in corridors and changing rooms, but not wet areas
Rethink	Kitchen mats and flooring

SHELF BIN KIT

PROPERTIES

Materials	Open-back, rolled-edge shelving with corrugated plastic storage bins
Features	Available in two heights 2000 and 1800mm
Capacity	Load bearing capacity 454kg
Supplier	Key Industrial & Commercial Equipment
Function	Factory storage
Rethink	Spice rack

CATERING STOOL

PROPERTIES

Materials	Plastic-coated steel frame; vinyl upholstered seat
Features	Available in three colours; stackable
Dimensions	450 x 700mm
Supplier	Key Industrial & Commercial Equipment
Function	Office equipment
Rethink	Kitchen stool

MATTING

PROPERTIES

Materials	Extremely resilient and flexible vinyl
Features	Double-sided, anti-slip, resistant to oil, grease, chemicals etc.; light, standard, economy and heavy duty
Dimensions	Available in 5 and 10m roll lengths
Supplier	Key Industrial & Commercial Equipment
Function	Industrial safety matting
Rethink	Kitchen flooring

DINING TABLE AND BENCHES

PROPERTIES

Materials	Steel frame, laminate frame
Features	Strong, versatile, light to handle, easy to store; available in four laminate colours; stackable benches, folding table
Dimensions	Table H698mm (three different lengths and widths); bench H432mm
Supplier	Gopak
Function	For school dining halls, factory canteens, youth organisations
Rethink	Kitchen/dining seating

SEATING UNIT

PROPERTIES

Materials	Laminate top, polypropylene seats with tweed or vinyl upholstery
Features	Modular units with matching tables
Dimensions	Table H720mm; seat H470mm
Supplier	Welconstruct
Function	For vending, refreshment and rest areas
Rethink	Kitchen table and seating

TRAFFIC CONE

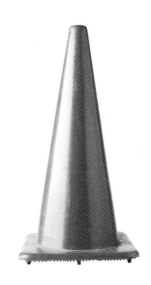

PROPERTIES

Materials	Super tough polymer
Features	Complete with ballast retaining skirt
Dimensions	H450, 530 or 730mm
Supplier	Signs & Labels
Function	Hazard warning and traffic control in commercial, industrial and leisure environments
Rethink	Kitchen no-go areas

STORAGE TROLLEY

PROPERTIES

Materials	Stainless steel trolley with lightweight fibreboard box trucks
Features	Double reinforced seams, 12mm external plywood base
Capacity	75, 100, 125kg
Supplier	Key Safety & Hygiene
Function	Factory storage
Rethink	Portable pantry or vegetable rack

RECYCLING BINS

PROPERTIES

Materials	LMPDE
Features	Separate compartments for bottles and cans
Capacity	78 or 156 litres
Supplier	Mailbox Healthcare Products
Function	Public waste bins
Rethink	Domestic recycling bins

MOBILE INGREDIENT TRUCKS

PROPERTIES

Materials	Rotationally moulded in heavy-duty food grade polyethylene
Features	Transparent flip top, shatterproof lid for easy identification of contents; available in four colours
Capacity	65, 89 or 132 litres
Supplier	Key Industrial and Commercial
Function	Dispensing and general handling of food ingredients in kitchens
Rethink	Domestic bulk-buy storage

FLAT MOP

PROPERTIES

Materials	Synthetic base with aluminium handle
Features	Mops faster than traditional mops; mop heads easily washed clean
Dimensions	H1100 x W839 x L540mm
Supplier	Bernard
Function	Industrial floor cleaner
Rethink	Kitchen mop

BENCH

PROPERTIES

Materials	Powder-coated finished steel; chipboard
Features	Easy to assemble: no nuts, bolts or bracing
Dimensions	H1574 x W1524 x D620mm
Supplier	Key Industrial & Commercial Equipment
Function	T-bar bench with drawer, louvred panel and top shelf for multi-purpose workstation
Rethink	Kitchen bench

DISPENSER

PROPERTIES

Materials	Plastic
Features	Blue wall-mounted dispenser with serrated centre area
Dimensions	H440 x W330 x D340mm
Supplier	Key Industrial & Commercial Equipment
Function	High-quality wiper dispenser to keep working environments clean and spill free
Rethink	Kitchen roll dispenser

ABSORBENT SOCK

PROPERTIES

Materials	Made from organic cellulose filler with tough outer skin
Features	High-bulk filler provides excellent floor hugging properties
Dimensions	3000mm long
Supplier	Key Industrial & Commercial Equipment
Function	Absorbs oils, coolants, solvents and water
Rethink	Draft excluder

SUPER TROLLEY

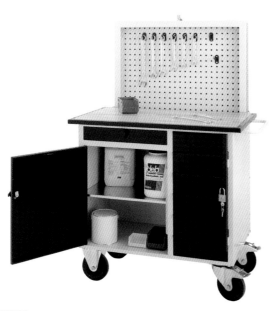

PROPERTIES

Materials	Heavy-duty welded steel construction; hardwood worktop
Features	Padlockable twist knob for security; comes with tool board
Dimensions	H929 x W1100 x 600mm
Supplier	Key Industrial & Commercial Equipment
Function	Maintenance trolley
Rethink	Kitchen workstation

MOBILE RACK

PROPERTIES

Materials	Polypropylene containers; epoxy powder-coated steel rack
Features	Supplied in component form for easy assembly; double-sided rack; colour-coded containers
Dimensions	H1065 or 1470 x W1000 x D500mm
Supplier	Welconstruct
Function	Free-standing storage rack
Rethink	Mobile kitchen pantry

RECYCLING BINS

PROPERTIES

Materials	Lightweight plastic
Features	Cobalt blue; supplied in multiple packs; European Community Regulation
Dimensions	Available in four sizes
Supplier	Signs & Labels
Function	Workstation recycling
Rethink	Kitchen recycling bin

FIRE EXTINGUISHER CABINET

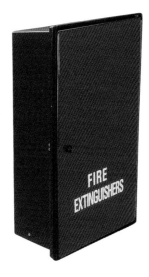

PROPERTIES

Materials	Glass-reinforced polyester resin
Features	Designed to house one or two extinguishers
Dimensions	Single: H840 x W325 x D240mm; Double: H840 x W540 x D240mm
Supplier	Signs & Labels
Function	Workplace safety
Rethink	Handy kitchen storage unit

MINI VALET

PROPERTIES

Materials	Rilsan-coated polystyrene tray
Features	Light, manoeuvrable, compact; two broom holder clips
Capacity	12 litre plastic bucket
Supplier	Bernard
Function	Cleaning and refuse collection
Rethink	Kitchen cleaning trolley

WALLPAPERING TABLE

PROPERTIES

Material	Plywood
Features	Fully foldable
Dimensions	Various
Supplier	Universally available at DIY suppliers
Function	Pasting wallpaper
Rethink	Desk

WORK

A wallpapering table — used as a desk — is an economic and versatile office solution. They are widely available in many formats, and always extremely lightweight, having been refined for on-site use by painters and decorators for the ease of room-to-room mobility.

The legs fold into the thickness of the table and the top folds to make a neat suitcase shape. All this, plus a handy printed measure on the work surface make this perfect modern office furniture.

Chilled items, such as live lobsters or ice-cream, get star treatment in thick-walled, styrene boxes that are generally disposed of after one journey. You do the environment and the retailer a favour by reusing them as scientific-looking storage.

Storage is often problematic and can be extremely expensive. However, using the food industry as a source of inspiration, storage can be completely free. With our ever-increasing interest in exotic foodstuffs, which have to be protected over long distances in a wide variety of transportation and handling systems, perishable and fragile foods now receive very sophisticated protection — perfect for use in other types of storage.

Engineering and ventilation suppliers carry a wide selection of attractive heavy-weight fans which knock the socks off the examples that are generally available through domestic outlets.

FREEZER BOX

PROPERTIES

Material	Expanded polystyrene
Features	Keeps hot food hot and cold food cold; portable with carrying strap
Dimensions	350 x 320 x 300mm
Supplier	Ekco Packaging
Function	Carrying food
Rethink	Stackable storage

WORK   Unchanged since Roman times, the humble house brick comes in endless variations. It is used in many ways, as a paperweight, doorstop, or to prop up a car while you steal the wheels. In this case (right), we see it in the more socially acceptable guise of a desk tidy.

Cable management is one of the big interior design problems of
the century. Every decade seems to bring a doubling of appliances
but what are we meant to do with all our hi-fi, answerphone,
ionizer, computer, scanner and printer plugs and leads?

Perhaps we are 10 or 20 years away from a cordless technology
that works with domestic appliances. In the meantime, we are
doomed unless we learn to love our plugs and cords.

Luckily they come in a variety of options. Wires come braided or
transparent, gold or fluorescent, fat, flat or coiled. Plugs can be
bought moulded onto the cord or colour-coded; they may be
round, square or transparent; then, plugged into an endless array
of adaptors, extensions or sockets.

Accept and even adore your cables, and the frustration of
untangling a mass of leads should disappear.

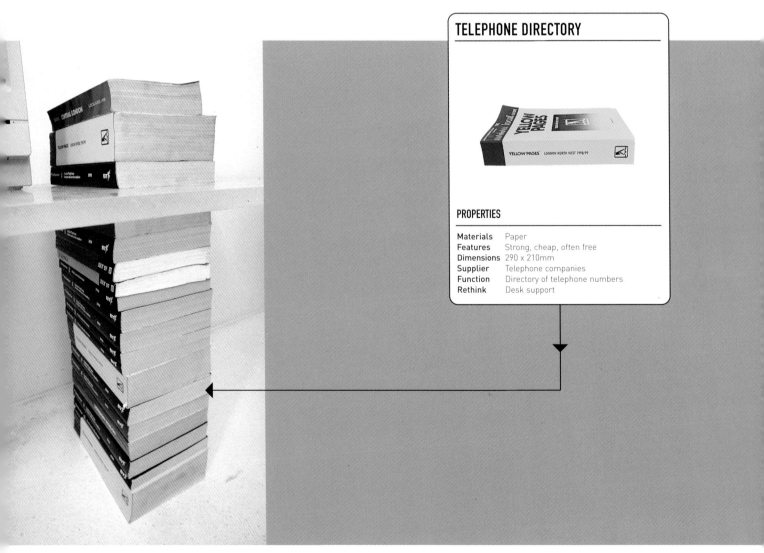

TELEPHONE DIRECTORY

PROPERTIES

Materials	Paper
Features	Strong, cheap, often free
Dimensions	290 x 210mm
Supplier	Telephone companies
Function	Directory of telephone numbers
Rethink	Desk support

 We are surrounded by artefacts that we seldom use but hoard, such as magazines and telephone directories. They mount in piles and
clutter valuable space until we get around to sorting them out. Why not put them to structural use as in the office desk solution above?

STOOLS

PROPERTIES

Materials	Polypropylene
Features	Fold away stools
Dimensions	Available in five sizes and nine colours
Supplier	Hille Furniture Ltd
Function	Classroom seating
Rethink	Seating for home office

MOBILE LIFT TABLE

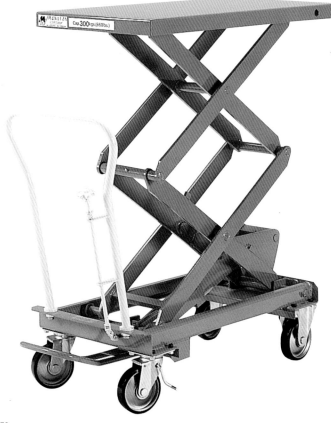

PROPERTIES

Materials	Robust steel construction; polyurethene-tyred wheels
Features	Manual hydraulic lift operated via foot pedal
Capacity	300kg medium duty table; H1563mm max
Supplier	Key Industrial & Commercial Equipment
Function	For lifting heavy work pieces – press tools, gearbox etc.
Rethink	Computer workstation

STORAGE CONTAINERS

PROPERTIES

Materials	Polyethylene, galvanized or painted steel
Features	Having identical height and base dimensions, can be interstacked
Dimensions	Available in multiple sizes
Supplier	Slingsby Commercial & Industrial
Function	Containers used in transportation
Rethink	Filing system

LITTER BIN

PROPERTIES

Material	Medium-density polyethylene
Features	UV stabilized, available in 10 colours, ground fixing post available
Capacity	40 litres
Supplier	Seton Industrial & Commercial Equipment
Function	Public waste bin
Rethink	Kitchen dustbin or recycling bottle bin

COLOURED TAPE

PROPERTIES

Materials	Colour-coded tape
Features	Pressure sensitive, available in 10 colours
Dimensions	L54m, W19mm, 38mm or 64mm
Supplier	Seton
Function	Used for colour banding with unprinted vinyl
Rethink	File marking

PLASTIC CONTAINER STORAGE SYSTEM

PROPERTIES

Materials	Injection-moulded polypropylene and stove-enamelled grey metalwork
Features	Colour-coded, wall-mounted storage system
Capacity	Boxes available in five sizes
Supplier	Welconstruct
Function	Small parts storage system
Rethink	Stationery storage

LIGHT FITTING

PROPERTIES

Material	Spun aluminium reflector
Features	Control gear housing construction allows high ambient temperature operation up to 45°C
Dimensions	W482 x H375mm
Supplier	Whitecroft Lighting Ltd
Function	Office lighting
Rethink	Home office lighting

CASTOR

PROPERTIES

Materials	Medium/light duty, pressed steel swivel; polypropylene and nylon wheel
Features	Braked castor, using foot operated pedal
Capacity	Loading capacity 125kg
Supplier	Key Industrial & Commercial Equipment
Function	Suitable for most applications
Rethink	Adding mobility to workstations

WAREHOUSE STORAGE

PROPERTIES

Materials	Steel-edged shelving, cardboard boxes
Features	6-bay shelf complete with 30 flip-top wallet boxes
Dimensions	Various
Supplier	Rapid Racking
Function	Warehouse filing system
Rethink	Home office filing

STORAGE BIN

PROPERTIES

Materials	Zintec-coated mild steel
Features	Lockable lid; liquid-tight sump complete with lifting handles
Capacity	50–1000 litres
Supplier	Signs & Labels
Function	Workplace storage of hazardous substances
Rethink	Office safe

MAINTENANCE TROLLEY

PROPERTIES

Materials	2mm-thick steep-plate worktop; stove-enamelled red
Features	Secure locking tool cabinet with one adjustable shelf
Dimensions	H610 x W450 x D450mm
Supplier	Slingsby Commercial & Industrial
Function	Heavy-duty maintenance trolley
Rethink	Computer station

FLOOR SIGN

PROPERTIES

Materials	3mm thick polyethylene
Features	Strong hinge, foldable for easy storage, impervious to harsh weather
Dimensions	305 x 508mm
Supplier	Seton
Function	Heavy-duty floorstand sign
Rethink	Desk trestle

CABLE DISPENSER

PROPERTIES

Materials	Fully welded tubular steel frame; epoxy powder-coated finish
Features	Can be used as a static unit or fitted with 1000mm diameter braked swivel castor
Dimensions	Available in two heights
Supplier	Key Industrial & Commercial Equipment
Function	For organized storage and dispensing of reeled products such as cable, rope and packaging materials
Rethink	Workstation

WHEELALONG STEP

PROPERTIES

Materials	Steel frame, rubber step
Features	Anti-slip tread; non-marking rubber tyred wheels for mobility
Dimensions	two-step H1205 x W455 x D605mm; three-step H1210 x W455 x D805mm
Supplier	Key Industrial & Commercial Equipment
Function	Ladder
Rethink	Library or kitchen steps

PLYWOOD CASE

PROPERTIES

Materials	4 or 5mm- thickness plywood
Features	No nails; prefabricated and sewn together with cold-rolled annealed steel; supplied in flat form for ease of storage
Dimensions	Available in 12 standard sizes; can also be manufactured to specification
Supplier	Key Industrial & Commercial Equipment
Function	For inland use and exporting
Rethink	Document storage

JUMBO BIN

PROPERTIES

Materials	Zinc-treated steel, epoxy powder-coated
Features	Three large litter openings; fire-resistant; available with chrome-plated ashtray recess
Capacity	60 litres
Supplier	Signs & Labels
Function	Large public areas, shopping centres, exhibitions, etc.
Rethink	Litter bin

PLASTIBOX STORAGE SYSTEM

PROPERTIES

Materials	Injection-moulded polypropylene and stove-enamelled grey metal work
Features	Self-stacking boxes
Capacity	Boxes supplied in five different sizes
Supplier	Welconstruct
Function	Clean for health, laboratory use, etc.
Rethink	Office storage

TOOL DRAWER TROLLEY

PROPERTIES

Materials	Welded tubular and flat steel frame
Features	Two lockable drawers; standard plate-fitted castors
Dimensions	H1030 x W600 x L1190mm
Supplier	Key Industrial & Commercial Equipment
Function	Robust trolley ideal for warehouse loads
Rethink	Computer trolley

TOOL TROLLEY

PROPERTIES

Materials	Heavy-duty sheet steel
Features	Six trays including one removable tote tray padlock facility
Dimensions	H267 x W571 x D424mm
Supplier	Key Industrial & Commercial Equipment
Function	Tool-box for professional mechanic
Rethink	Small equipment/stationery store

GRIT BIN

PROPERTIES

Materials	Vandal- and UV-resistant polyethylene
Features	Large surface area and double-skinned lid
Capacity	7 or 14 cubic foot
Supplier	Slingsby Commercial & Industrial Equipment
Function	Used on highways or private premises
Rethink	Toy tidy

PLAY

Some types of storage never come big enough for their intended use. The laundry bin is a good example. For anybody with a family, an interest in sport or a compulsive hygiene obsession, a domestic laundry bin just doesn't do the job. In a similar fashion, toy storage

is frequently undersized for the contemporary over-consuming child. Industry, however, is expert at solving large storage problems. The grit bin is a perfect example of a design perfected for hostile environments, and is equally well suited to the challenges of a kid's room.

Greenhouses are miniature masterpieces in modern architecture. Their modular lightweight construction lends itself to rapid assembly and they are phenomenally strong despite their puny appearance. They can be used internally as temporary spaces to isolate oneself from the hurly burly of family life, or to isolate children's mess and noise from the rest of the house. However, for safety it is essential to use polycarbonate glazing in such situations.

Children love the Wendy-house appearance of the greenhouse, especially when it is glazed with translucent coloured plastic.

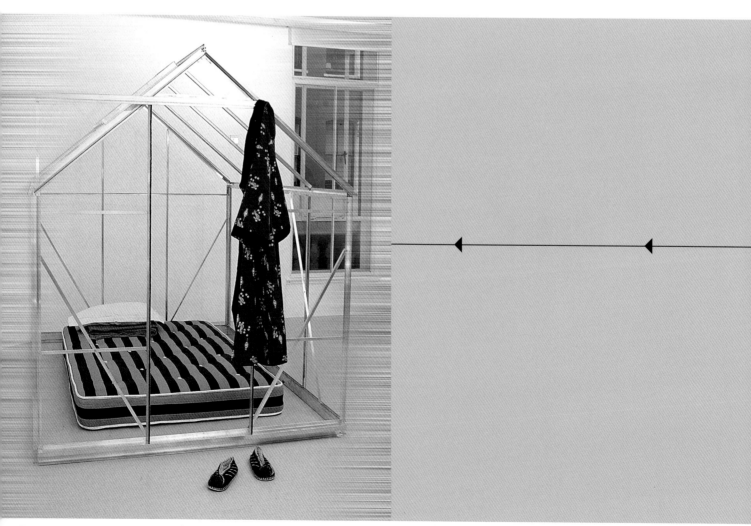

Children and greenhouses may seem a recipe for disaster, but not when traditional glass is replaced with vandal-proof polycarbonate. The lightweight structure makes them excellent playhouses, or even a place for adults to take refuge.

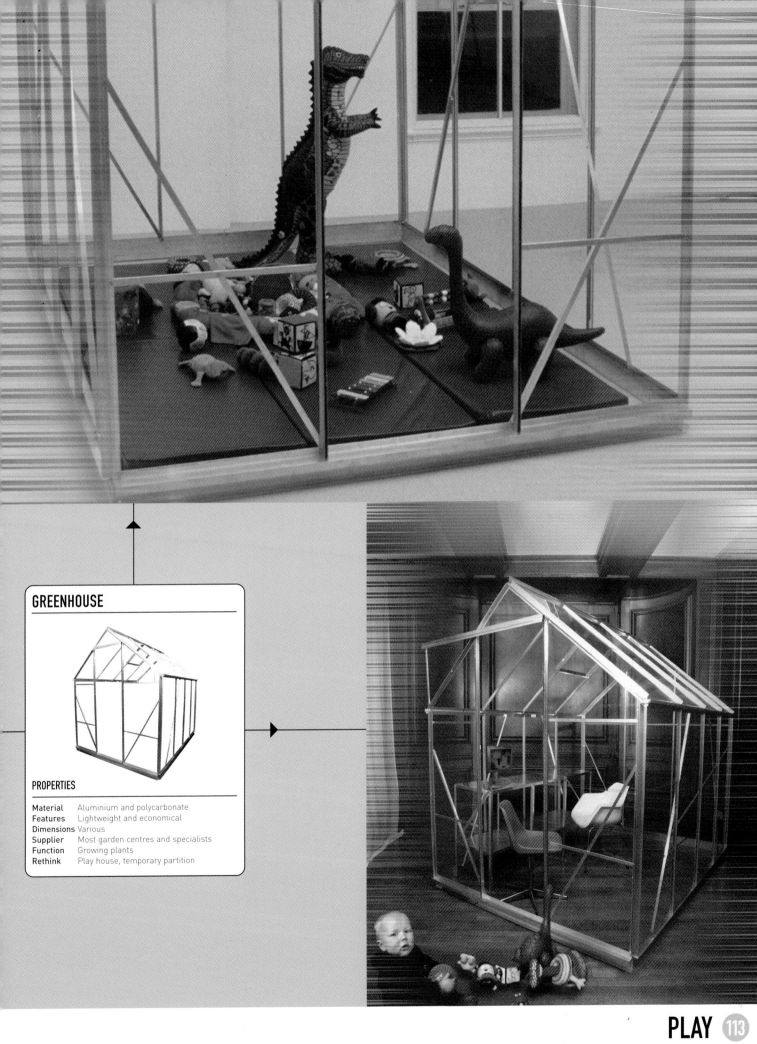

GREENHOUSE

PROPERTIES

Material	Aluminium and polycarbonate
Features	Lightweight and economical
Dimensions	Various
Supplier	Most garden centres and specialists
Function	Growing plants
Rethink	Play house, temporary partition

PLAY All kinds of businesses use temporary buildings for shelter, security or privacy, and these can often be used for other purposes. In this case, a telephone engineer's tent — normally used to shelter workers in holes in the ground — is being used as a Wendy house.

In a relentlessly commercial world, children ought to be encouraged to think of others, but may need incentives. These large figurative charity collection boxes are usually found outside high street shops, but they are an asset to any child's bedroom, and should entice youngsters to give generously and regularly, to charity. Illustrated is the Mencap blue bear, but there are endless variations, including dogs, Wombles and Noddy.

Furniture used in infant schools has the dual advantages of high specification and low cost. They also have the added benefit of being workmanlike miniatures of office furniture, and therefore seem more grown-up to kids than the average domestic versions. Safety tested and easy-to-clean features come as standard — evidently essential where children are concerned.

INFANT SCHOOL CHAIR

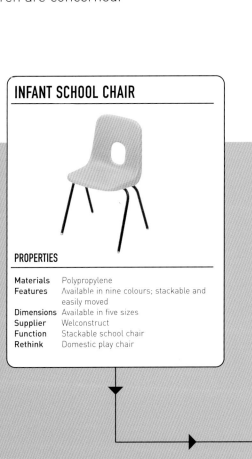

PROPERTIES

Materials	Polypropylene
Features	Available in nine colours; stackable and easily moved
Dimensions	Available in five sizes
Supplier	Welconstruct
Function	Stackable school chair
Rethink	Domestic play chair

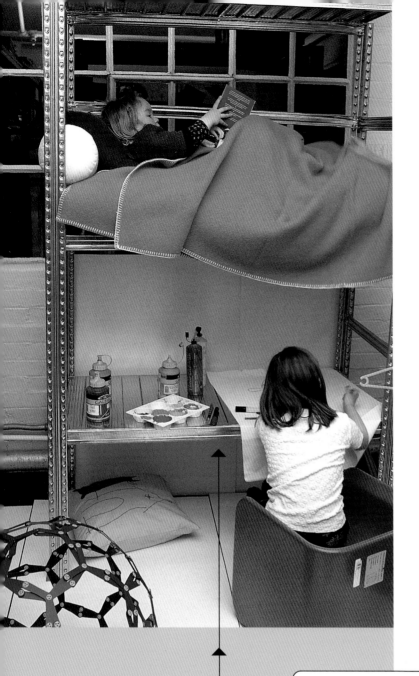

The modular nature of these types of racking systems provides an ideal solution to many everyday construction problems. They are both easy to install and adapt as the family expands and grows up, but would be equally suited to home office or kitchen storage. Heavy-duty structural systems are also available.

Used by industry for all types of load-bearing applications, they can be adapted in the home to create balconies, libraries or mezzanine floors.

MELAMINE SHELVING

PROPERTIES

Materials	Steel-edged melamine shelving
Features	No nuts and bolts; each shelf supports four adults; available in 7 colours
Dimensions	H1980 x W915 or 1220mm
Supplier	Rapid Racking
Function	Office or retail shelving
Rethink	Toy storage

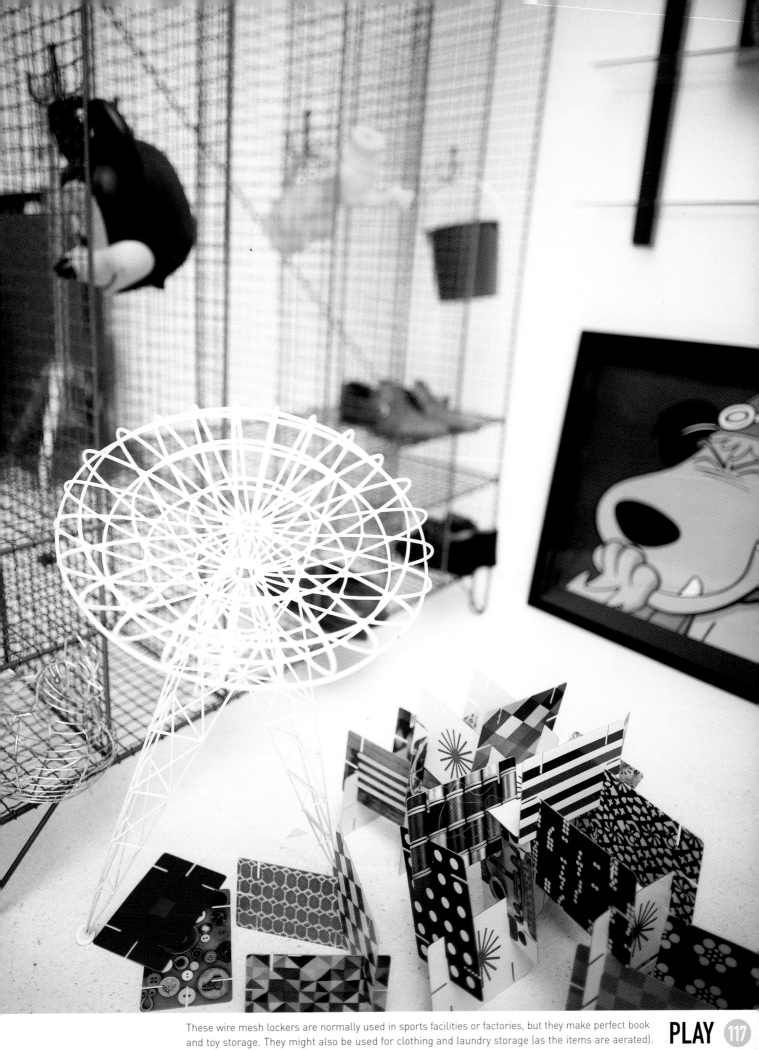

These wire mesh lockers are normally used in sports facilities or factories, but they make perfect book and toy storage. They might also be used for clothing and laundry storage (as the items are aerated).

RUBBISH BIN

PROPERTIES

Materials	High-impact resistant polyethylene
Features	Ultra violet stabilized to withstand the toughest environmental conditions
Capacity	84, 120, 155 litres
Supplier	Slingsby Commercial & Industrial Equipment
Function	Hooded litter bin, for use on garage forecourts
Rethink	Small toy storage

LOCKERS

PROPERTIES

Materials	Tough durable plastic
Features	Wide variety of configurations can be assembled to utilize available space
Dimensions	Available in two sizes H305 or 915 x W305 x D460mm
Supplier	Welconstruct
Function	Suitable for leisure, educational, industrial or commercial use
Rethink	Storage, room dividers

MOBILE BIN

PROPERTIES

Materials	High density polyethylene
Features	Resistant to ultra violet light and extreme weather conditions; available in British Racing Green, grey, blue, yellow, burgundy and green
Capacity	90 - 1100 litres
Supplier	Key Safety & Hygiene Equipment
Function	Mobile refuse containers
Rethink	Room tidy

TROLLEY

PROPERTIES

Materials	Polypropylene containers; solid rubber-tyred swivel castors
Features	Can be used with either 4 small containers, 2 large containers or a combination of both
Dimensions	L1010 x W570 x H990mm
Supplier	Slingsby Commercial & Industrial Equipment
Function	Order collation and counter fill-up
Rethink	Water and sand play trays or toy storage

VINYL TAPE

PROPERTIES

Materials	Durable pressure-sensitive vinyl
Features	Over-laminated for extra protection against chemicals and weathering
Dimensions	W51mm and 102mm x L16.5 or 33m
Supplier	Seton
Function	Tape to indicate flow direction of pipe contents
Rethink	Nursery wall decoration

ILLUMINATED SIGN BOX

PROPERTIES

Materials	Powder-coated aluminium sign box; 3mm acrylic substrate sign face
Features	Design-a-sign service — choose your own text and typeface
Dimensions	H458 x W625mm
Supplier	Signs & Labels
Function	Ideal for highlighting retail or commercial premises
Rethink	Child's room sign

KIK STEP

PROPERTIES

Materials	Steel construction
Features	Retractable spring-loaded castors; available in six colours
Dimensions	H355 or 405mm
Supplier	Welconstruct
Function	Office accessory
Rethink	Children's seating and steps

JUNIOR TABLE AND CHAIRS

PROPERTIES

Materials	Decamel laminate table top with PVC edging; co-polymer polypropylene moulded seat
Features	Choose table height to match junior chairs
Dimensions	Made according to age group – from 3–5 to 13+ years
Supplier	Key Industrial & Commercial Equipment
Function	Ideal for schools colleges, training rooms, etc.
Rethink	Child's table and chairs

CONTAINERS

PROPERTIES

Materials	Injection-moulded polypropylene
Features	Self-stacking, choice of five bright colours
Dimensions	Available in five different sizes
Supplier	Welconstruct
Function	Office storage system
Rethink	Toy storage

NO WAITING CONE

PROPERTIES

Materials	Super tough polymer, tested to -16°C
Features	Raised moulding in regulation blue and red colours
Weight	Approx 3.5kg
Supplier	Signs & Labels
Function	Traffic cone control
Rethink	No-go areas and imaginative play prop

PALLET

PROPERTIES

Materials	High-density polyethylene
Features	Open-ribbed deck pallet; bears loads up to 4000kg
Dimensions	800 x 1200mm
Supplier	Slingsby Commercial & Industrial Equipment
Function	Pallet for food manufacturers and meat processors
Rethink	Bed base, play area

INFANT CHAIR

PROPERTIES

Materials	Polypropylene
Features	Available in nine colours; stackable and easily moved
Dimensions	Available in five sizes
Supplier	Hille Ltd
Function	School seating
Rethink	Domestic seating

SAFETY STORE

PROPERTIES

Materials	Polyethylene
Features	Removable floor grid for easy cleaning
Dimensions	L1615 x W1615 x H1870mm
Supplier	Welconstruct
Function	Hazardous material safety store with entrance ramp
Rethink	Garden playhouse

STEP LADDERS

PROPERTIES

Materials	Tubular steel frame
Features	Wide non-slip tread; concealed spring-loaded castors
Dimensions	With 2, 3, 4, or 5 steps
Supplier	Welconstruct
Function	Library, office, factory access steps
Rethink	Bunk-bed ladder

MELAMINE SHELVING

PROPERTIES

Materials	Steel-edged melamine shelving
Features	No nuts and bolts; each shelf supports four adults; available in 7 colours
Dimensions	H1980 x W915/ 1220mm
Supplier	Rapid Racking
Function	Office or retail shelving
Rethink	Toy and clothes storage

CHIPBOARD SHELVING

1400 LBS

PROPERTIES

Materials	Chipboard shelves, steel edging
Features	Assembles in seconds, fully adjustable
Dimensions	H1980 x W915/1524mm
Supplier	Rapid Racking
Function	Warehouse storage
Rethink	Bunk-bed/tough storage for toys

SEATING

PROPERTIES

Materials	Polypropylene
Features	Stackable in nine colours and five sizes
Supplier	Hille Ltd
Function	Classroom seating
Rethink	Playroom furniture

SAFETY BARRIER

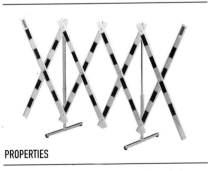

PROPERTIES

Materials	UPVC white arms; galvanized tubular steel legs and feet
Features	Free standing; concertina-extending-style barrier, fitted with 32 reflectors
Dimensions	H1000 x W2200mm (extended)
Supplier	Welconstruct
Function	Protection of pedestrians, machinery, etc.
Rethink	Toddler safety barrier

BARRIER PLANK

PROPERTIES

Materials	UPVC planks with high-impact steel posts and rigid moulded thermoplastic base
Features	Easy to assemble, excellent stability
Dimensions	H1200 x L2000mm
Supplier	Signs & Labels
Function	Site traffic management system
Rethink	No-go areas

BUY

ABE Catering Equipment
ABE Projects
Head Office
Holloway Head
Birmingham B1 1NU
Tel: 0121 622 6288
Fax: 0121 622 7229

Bernard Cleaning,
Maintenance & Hygiene
Bernard Supplies Ltd
PO Box 207
Ashford
Kent TN24 8ZG
Tel: 01233 640400
Fax: 01233 640700
Internet: www.bernard.com

Bristol Maid
Hospital Metalcraft Limited
Blandford Heights
Blandford Forum
Dorset DT11 7TE
Tel: 01258 451338
Fax: 01258 455056

Display Lines
Pondcourt
Loxwood
Billingshurst
West Sussex RH14 0SA
Tel: 01403 752686

Ekco Packaging
Asheridge Road
Chesham
Buckinghamshire HP5 2QF
Tel: 01494 775221

Gopak Limited
Range Road
Hythe
Kent CT21 6HG
Tel: 01303 265751
Fax: 01303 268282
Internet: http://www.gopak.co.uk/
email: gopak@gopak.co.uk

Hille Limited
Cross Street
Darwen
Lancashire BB3 2PW
Tel: 01254 778850
Fax: 01254 778860

Key Industrial
& Commercial Equipment
Blackmoor Road
Ebblake Industrial Estate
Verwood
Dorset BH31 6AT
Tel: 01202 825371
Fax: 01202 813348

Key Safety & Hygiene
Blackmoor Road
Ebblake Industrial Estate
Verwood
Dorset BH31 6AT
Tel: 0990 761761
Fax: 01202 826453

Mailbox Healthcare Division
Mailbox International Limited
Bayley Street
Stalybridge
Cheshire SK15 1QQ
Tel: 0161 3305577
Fax: 0161 3305576

Philip Harris Scientific
Novara House
Excelsior Road
Ashby Park
Ashby de la Zouch
Leicestershire LE65 1NG
Tel: 0845 6040490
Fax: 01530 419300

Pottery Crafts Limited
Sales Showrooms
Campbell Road
Stoke-on-Trent
Staffordshire ST4 4ET
Tel: 01782 745000
Fax: 01782 746000

Rapid Racking Limited
Kemble Business Park
Kemble
Cirencester
Gloucestershire GL7 6BQ
Tel: 01285 686800
Fax: 01285 686900
Internet: www.rapid-racking.co.uk

Seton Identification & Safety
Buyer's Guide
Seton Limited
Dept AQ
PO Box 77
Banbury
Oxon OX16 7LS
Tel: 0800 585501

Seton First Aid Catalogue
Seton Limited
Dept AV
PO Box 77
Banbury
Oxon OX16 7LS
Tel: 0800 525042

Signs & Labels Limited
Premises Management Catalogue
Bredbury Industrial Park
Stockport
Cheshire SK6 2SD
Tel: 0161 494 6125
Fax: 0161 430 8514
Internet: www.signsandlabels.co.uk

Slingsby Commercial & Industrial
Equipment
HC Slingsby plc
Preston Street
Bradford BD7 1JF
Tel: 01274 721591
Fax: 01274 723044
Internet: www.slingsby.com
email: sales@slingsby.com

Thetford Limited
Centrovell Industrial Estate
Caldwell Road
Nuneaton
Warwickshire CV11 4UD
Tel: 01203 341941
Fax: 01203 641230

Welconstruct
127 Hagley Road
Edgbaston
Birmingham B16 8XU
Tel: 0121 4559798
Fax: 0121 4548114

Whitecroft Lighting Limited
Central Office
Burlington Street
Ashton-under-Lyne
Lancashire OL7 0AX
Tel: 0990 087087
Fax: 0990 084210
email@whitecroftlight.com

LIST

FINISH

Acknowledgements
The publisher would like to thank the following
photographers and organisations for their kind permission
to reproduce the photographs in this book:

8 Tim Hall/Robert Harding Picture Library;
9 Steffan Hill/Frank Spooner Pictures;10 above Colin
Dixon/Arcaid;10 below Caroline Penn/Impact;12 David
Lees/Corbis;13 Mitsuhiro Wada/Liaison International;
14 Tony Arruza/Corbis;14-15 Liz Hymans/Corbis;16 left
Buddy Mays/Corbis;16 right H.Rogers/Trip;17 H.Rogers/
Trip;18 Hans Van der Mars(designer:Rody Graumans/Droog
Design);19 left courtesy Tate Gallery, London, (artist:Carl
Andre)/DACS;19 right Stephen White courtesy Jay
Jopling/White Cube, London (artist: Damien Hirst);20 left

Hideyo Kubota; 20-21 John Davies courtesy Cardiff Bay
Development Corporation (artist: Pierre Vivant); 22 Cary
Wolinsky/Trillium Studios; ;86 Mike McQueen/Impact

The publisher would also like to thank the following
suppliers and manufacturers for supplying us with images
of their products:

Bernard Suppliers, Bisley Office Equipment, Bristol Maid,
Gopak Ltd, Hille (Idem Furniture Ltd.), Kartell, Key
Industrial Equipment Ltd, Pottery Crafts Ltd., Rapid
Racking, Seton Ltd., Signs and Labels Ltd., Slingsby
Commercial & Industrial Equipment, The Stamford Group,
Thetford, The Welconstruct Company, Whitecroft Lighting

Every effort has been made to trace the copyright holders
and we apologise in advance for any unintentional omission
and would be pleased to insert the appropriate
acknowledgement in any subsequent edition.

Thanks to
Harriet Devoy of Johnson Banks; Hilary Burden;Sue Seddon

and to
Armitage Shanks; Display Lines; Eden Greenhouses; Ekco
Packaging; Philip Harris Scientific; Mailbox Mouldings;
Seton Ltd and HC Slingsby plc